T0251585

Ergonomics Analysis
and
Problem Solving
Manual

Mike Burke

CRC Press
Taylor & Francis Group
Boca Raton London New York

CRC Press is an imprint of the
Taylor & Francis Group, an **informa** business

First published 1993 by CRC Press
Taylor & Francis Group
6000 Broken Sound Parkway NW, Suite 300
Boca Raton, FL 33487-2742

Reissued 2018 by CRC Press

Library of Congress Cataloging-in-Publication Data

Catalog information is available from the Library of Congress.

ISBN 0-87371-651-5

A Library of Congress record exists under LC control number: 94790371

Publisher's Note
The publisher has gone to great lengths to ensure the quality of this reprint but points out that some imperfections in the original copies may be apparent.

Disclaimer
The publisher has made every effort to trace copyright holders and welcomes correspondence from those they have been unable to contact.

ISBN 13: 978-1-315-89279-5 (hbk)
ISBN 13: 978-1-351-07189-5 (ebk)

Visit the Taylor & Francis Web site at http://www.taylorandfrancis.com and the
CRC Press Web site at http://www.crcpress.com

Congratulations on your purchase of *Ergonomics Analysis and Problem Solving!*

I am certain that with a little investment in time now, you will save hours of time and effort down the road. This program has been developing for more years than I care to recall at the moment. MCI made a lot of money in the lengthy phone conversations between myself and Sherm the programmer.

To assist in the evolution of this program, I am asking that users provide feedback and suggestions. The answers to any questions as well as resolutions of any problems can be shared with anyone using this program.

In order to accomplish this please fill out and send in the information sheet to join the *Ergonomics Analysis & Problem Solving* Users Group. I promise that your name will not be sold to any mailing lists. I will send out to all members an occasional letter which will discuss issues and experiences in using this program.

Name _____

Address _____

_____

Phone (Optional) _____

Fax (Suggested) _____

Date Program Purchased _____

Send the Information above to:

Mike Burke
Ergonomix Inc.
P.O. Box 753
Lake Zurich, Illinois 60047

ERGONOMICS ANALYSIS AND PROBLEM SOLVING OPERATING MANUAL

TABLE OF CONTENTS

INTRODUCTION

Organizations and companies are trying to use the science of ergonomics to help decrease the incidence of musculoskeletal injuries and illness by redesigning or retrofitting their work stations. This program will deal with the issue of controlling cumulative trauma disorders by focusing on the mechanical stresses in the workplace. It should be understood that while your efforts may be driven by the need to control cumulative trauma disorders, maximum productivity and consistent quality will surely result.

In many cases the number of work stations or processes in need of ergonomic analysis within a facility is very large. It may be considered financially unfeasible or impractical for a professional ergonomist to provide comprehensive services to an organization or company. Many companies have turned to in-house individuals or committees to perform this needed service. In some cases this is an adjunct to having a professional ergonomist deal with the most challenging ergonomic situations.

There is a great deal of self analysis that can be performed by individuals with an understanding of the processes, products, and politics of their particular facility. In many cases, this type of autonomous ergonomics management is even more effective than attempting to use a professional ergonomist.

Of course, there may be situations where only an outside consultant will be able to provide the necessary service. Use of these initial steps will help to identify where those situations arise.

This program is a tool to help individuals with limited ergonomics experience to consistently perform work station analysis. It assists in the production of an ergonomic jobsite analysis report. The operator enters specific information about the job being studied. This specific information will be integrated into several interim reports as well as a final comprehensive report.

Overview Of The Program

The goal of an ergonomic analysis is to identify conditions in a work station which may act as barriers to maximum productivity, obstacles to consistent quality, and challenges to immediate worker comfort and long-term employee well being. Once this identification process has been completed, various strategies to promote an efficient and safe work environment are determined.

This process involves three separate phases. The phases are background information gathering, ergonomic risk factor identification, and determination of effective interventions.

Compiling Background Information

In the Background information phase, the job is broken down into its component duties and tasks. Each task can then be quantified by the rate of performance, the number of hours that it is performed, and the total number of times per day that it is performed. This provides a simple framework made up of definable units of time for the next step, which is the identification of ergonomic risk factors.

The Ergonomics Analysis and Problem Solving software program takes the operator through a series of screens where fields are filled in to determine a duty list, a task list, and a task quantification list. Simple arithmetic operations such as ongoing calculation of percentages for duty and task exposure help the novice and professional alike. Context-sensitive help screens are available for each step along the way to define terminology and explain how to get around the screens.

Following completion of the Background Information phase, the program can generate a series of reports which can list each task, designate a unit of completion for each task, indicate the time required to perform a single unit of completion, present total time that a target worker performs a task, and finally, state the number of times a task is performed each day.

Identification of Ergonomics Risk Factors

An ergonomic risk factor is a condition or practice in a work station which may act as a barrier to maximum productivity, an obstacle to consistent quality, or a challenge to immediate worker comfort or long term employee well being. These risk factors can be classified into five basic categories. These categories are end range positions, unsupported postures, forceful exertions, environmental factors, and excessive energy demands. Definitions of these terms cam be found in the Risk Identification section of this manual.

This phase of the program uses a grid format, allowing a cursor to move freely throughout the grid. Risk factors are listed along the top of the grid and regions of the body are listed along the vertical axis. As the operator moves the cursor though out the grid, a general statement is displayed. This statement consists of the risk factor and the body part. When the operator identifies a risk factor, he positions the cursor at the appropriate position on the grid and hits the <En-

ter> key. This brings up a series of menus to help further define this risk factor. The details are added to the general statement and is displayed throughout the process.

When a risk factor is identified, then that factor is quantified by the appropriate details and additional information. This may include body part position, amount of times it occurs per unit of completion for that task, type of force, amount of force, type of environmental factors or amount of time of environmental exposure.

This product of this phase is a list of consistently stated "ergonomic risk factors" and the corresponding job, duty, or task where they occur.

The program offers the operator the option to perform the risk identification without performing the complete task breakdown in the Background Information stage; however this compromise will require the operator to type in quantity information manually.

Determination of Ergonomic Interventions

The final phase of the ergonomics analysis process is the determination of effective interventions. In order to insure that as many interventions as possible are considered, several different approaches should be pursued. These approaches can be classified into three different categories. These categories are process modifications, worker involved interventions, and work station redesign or retrofitting.

Process modification looks at the input and the output of the process. The analyst focuses on the input of the process and consider how it can be changed to reduce or eliminate the risk factors identified . Typical considerations here are the position of the input as presented, the rate at which it is presented, the size, shape, weight, and even temperature of the input. The same steps are then taken for the output of the process. This may involve adjustments to the rate of the output or even the level of completion of the product.

Worker involved interventions are those which require participation on the part of the worker. Possibilities in this area include education programs, exercise programs, personal protective equipment, job rotation or light duty programs, or even mandating procedures for the performance of a task.

Workstation redesign or retrofitting includes all the objects that make up the workstation environment. This would include the work station itself as well as the tools, carts, environmental conditions, and many other items. The workstation typically lends itself to the greatest number of possible interventions.

The program provides various possible interventions from which to choose. These interventions are organized by categories which fall into the three approaches presented above. The program provides several prompts to encourage entering original ideas.

Process Based Interventions
    Input
    Output

Worker Based Interventions
    Training Format Options
    Training Content Options
    Exercise Content Options
    Exercise Format Options
    Rotation
    Personal Protective Equipment
    Mandate Procedures
    Worker Selection
    Monitoring Devices
    Other

Workstation Based Programs
    Movement Assistance
    Assistance To Hold In Place
    Workstation Adjustments
    Worker Support
    Tool Considerations
    Maintenance
    Information Display
    Control Considerations
    Facilitate Access
    Decrease Repetitions
    Environmental Controls

The various categories are displayed along with examples of interventions. The operator can look over the interventions listed to choose the appropriate ones for their unique work environment. The idea is not to provide every possible type of intervention but rather to help foster the development of new and innovative ideas without overlooking ones that are more obvious and easy to implement.

Uses For This Program

Part of a Comprehensive Process

A comprehensive ergonomics program would involve:

1. Determining ergonomic needs in specific areas,
2. Determining the demands of the targeted job,
3. Performing the analyses of those targeted areas,
4. Discovering various interventions to decrease or eliminate potentially hazardous conditions,
5. Screening those interventions,
6. Implementing the interventions, and
7. Tracking the effectiveness of those interventions.

This program provides the tools for the performance of steps 2, 3, and 4 in the above model. It is the result of years of job site analysis experience using various techniques to find the procedures which are the least cumbersome, most practical, and objective. As there is likely to be a great deal of variability in the situations you encounter, it is meant to be as flexible as possible. At almost any point in the program you may record "additional comments" which may not fit into the protocol presented but are nevertheless important.

Work Station Design

The most likely use of this program will be to perform an ergonomics analysis of an existing workstation. It can also be used to project the likely ergonomic hazards of a prototype workstation.
Americans with Disabilities Act Compliance
This program can be extremely helpful in determining the specific tasks that comprise a job. This can provide an objective job description in the form a quantitative task list. In addition, the Intervention Discovery process can provide a basis for reducing unnecessary demands in order to accommodate a greatest number of job applicants.

Persons Likely To Benefit By Using This Program

Anyone with a sincere interest should be able to use this program to identify and solve most of the conditions likely to be present in a work station. It is particularly effective for individuals or groups with a limited amount of time to focus on ergonomics analysis.

This program would meet needs of the following groups:

> In-house Ergonomics Committees
> In-house Health and Safety Professionals
> Occupational Health Providers
> Risk Managers
> Human Factors Specialists
> Professional Ergonomists

HOW TO USE THE PROGRAM

An ergonomic jobsite analysis is a process which involves observation of a worker performing all or some of their job. Each job can be broken down into units called Duties. Each duty can be broken down into tasks. These terms will be defined later in this manual. The amount of time and or the number of times that each duty or task is performed is listed.

Specific potentially hazardous ergonomic conditions (Risk Factors) are identified by the analyst. The approach is to look at specific regions of the body and identify specific positions, postures, forceful exertions, and environmental exposures. The frequency of the risk factor or the amount of exposure is determined based on the information determined in the Duty and Task lists.

Suggestions (Ergonomic Interventions) for alleviating or reducing risk factors and therefore making the job safer are determined. Interventions in the form of modifications or additions to the workplace, changes to the procedures, or employee participation in specific programs are listed.

Finally, a customized report reflecting the findings of the analysis can be generated.

Help!

Help is available at any time by striking the <F1> Key

Getting Started

From the C:\ prompt,

1.   Type CD\ERGO. The screen will display c:\ERGO
2.   Type ERGO. The screen will display a title screen for a few seconds and then display the Main Menu.

The Main Menu will offer you the opportunity to select which phase of the jobsite analysis processes you wish to perform. The first step will always be to enter a new job or recall a previously entered

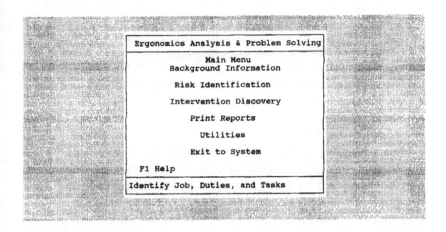

one. To do this you must first choose the "Background Information" option.

1.   Use the arrow keys to highlight "Background Information" and strike <Enter>.

The "Background Information" menu will be displayed. There are several steps in the background information process. These steps are displayed in the menu below.

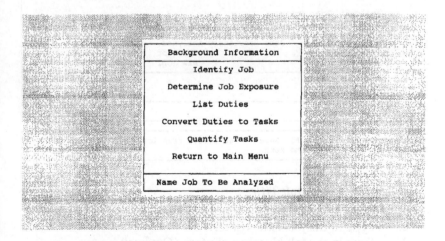

Background Information

The goal of this step is to produce a description of the job to be analyzed in the most practical units. This will facilitate the risk identification process by providing clearly definable units upon which to focus and quantify risk factors. In addition this will allow you to be certain that the job you are analyzing from year to year has remained the same. The final product here is a list of duties or tasks and the amount of time or the number of repetitions for each. Each duty or task can then be evaluated individually.

How to Identify Job

This screen can be accessed by highlighting "Identify Job" from the Background Information menu and striking <Enter>. The cursor will be displayed in the "Company "space. Either type in a new company name and job or select one from jobs already entered. Entering this information will allow you to save all your data about this job.

The program will not allow you to continue until you have entered a company name and job title or selected one from those already entered. Entering a job number is opional.

```
|                          Identify Job                              |
|                                                                    |
|    Company:                                                        |
|   Job Title:                                                       |
|  Job Number:        0                                              |
|     Analyst:   Mike Burke                                          |
|        Date:   10-04-92                                            |
|                  ---- Jobs Previously Analysed ----                |
|      Ergonomix Inc                  Operator              0007     |
|      NEW JOB 1                      NEW JOB1              0004      |
|                                                                    |
|--------------------------------------------------------------------|
| Enter company name                                                 |
|--------------------------------------------------------------------|
| F1 Help     F3 New     F10 Save    Tab Select From List   PgDn Next|
| Esc Return to Background Menu                                       |
```

How to Add a New Job Title
1.   Type in the Company Name <Enter>
2.   Type in the Job Title <Enter>
3.   Type in the Job Number <Enter> (*Optional*)
4.   Strike <Page Down> to save this information and proceed to the next step.

How to Select a Previously Entered Job

1. Strike the <Tab> Key to move the cursor into the menu box.
2. Use the Up and Down Arrow keys to highlight the desired job title.
3. Strike <Enter> to select and move on to "Determine Job Exposure" step. The "Determine Job Exposure" screen will be displayed.
4. If the job identification information displayed at the top of the screen is not correct, strike <Escape> to return to the Background Information Menu.

How to Modify "Jobs Previously Analyzed" Identification Data

1. Strike the <Tab> Key to move the cursor into the menu box.
2. Use the Up and Down Arrow keys to highlight the desired job title.
3. Strike <Tab> to move back into the main screen to make modifications to information.
4. Strike <Page Down> to proceed to the "Determine Job Exposure" step.

How to Determine Job Exposure

This screen can be accessed by either selecting it from the Background Information Menu, or by striking <Page Down> from the Identify Job screen. The screen is already set up to reflect an eight hour job exposure. You may accept them as is or, use steps detailed below to determine a different job exposure.

```
                        Determine Job Exposure
  Company: Sample Company                                    11-4-96
  Job Title: 12345                              Job Number:    678


                    Enter on duty time:       7:00 AM
       Enter number of minutes for first break:  15
          Enter number of minutes for lunch:     30
       Enter number of minutes for second break: 15
  Enter number of minutes any other 'Down' time:  0
                    Enter off duty time:      3:30 PM
       Enter number of days/shifts per week:     5
          Enter number of weeks per year:       48

                    Daily Job Exposure 7.5   hrs

  Enter on-duty hour (1-12)


  F1 Help     F9 Default    F10 Save    PgUp Previous    PgDn Next
  Esc Return to Main Menu
```

The purpose of this step is to determine the amount of time that a worker actually spends performing work. In most cases this may be different from either the number of hours spent at work or the number of hours that a worker is paid for.

1. Type in the Time that a worker goes on Duty including AM or PM <Enter>
2. Type in the number of minutes of the first coffee break of the day <Enter>
3. Type in the number of minutes for a mid shift meal break. (If there are several meal breaks, then add up the total number of meal break minutes and type in here) <Enter>
4. Type in the number of minutes of the second coffee break of the day <Enter>
5. Type in any additional down time. This includes any known set up times, unofficial breaks, or anticipated unofficial rest periods (goofing off time). <Enter>
    The Arrow up and Arrow down keys can also be used to move around in this screen.
6. Strike <Page Down> to save this information and proceed to the next step.

How to List Duties

This screen can be accessed by either selecting it from the Background Information Menu, or by striking <Page Down> from the Job Exposure screen.

```
                              List Duties                    11-4-96
     Sample Company
     Title - 12345                                    Job Number - 678

                     Total Job Exposure  7.50 Hours

         Duty #                    Description          EXPO      %
            1                                              0       0
            2                                              0       0
            3                                              0       0
            4                                              0       0
            5                                              0       0
            6                                              0       0
            7                                              0       0
            8                                              0       0
            9                                              0       0
           10                                              0       0
               Other Unspecified Duties                  7.5     100


     F1 Help    F3 Add    F4 Delete    F10 Save    PgUp Previous
     PgDn Next    Esc Abort
```

In this step you will create a list of duties for this job title. A Duty is a general statement which describes an activity which is a responsibility of a worker. Examples would be statements such as, Loading Trucks, Cleaning up area, or Performing paperwork. In addition to listing the duties, you are also given the opportunity to type in the "duty exposure" or number of hours that each duty is performed.

What if the duties vary from day to day?

It may be difficult to arrive at a "typical day", In these cases you should continue the analysis using a given day and one example of how much of each duty is performed. Once you have saved this information, you can change the job identification numbers and repeat the process to reflect a different day. It may be helpful to perform a series of "typical days " which will reflect the extreme of exposure to each duty.

Modifications to previously entered information can be made an any time. It may be necessary to estimate the number of hours that a duty is performed each day.

1. Type in a description of a job duty <Enter>
2. Type in the number of hours or decimal fractions that the corresponding duty is performed <Enter>
3. As the exposure in entered, it will be deducted from the "Other unspecified duties" displayed at the bottom of the screen.
4. Continue to enter duties until all are entered.

An alternate method for moving around this screen is to strike the <Tab> key to move from the "Description Column" to the "Exposure Column". To move from the Exposure Column to the description Column you will need to Hold down the Shift Key and strike <Tab>.

Strike <Page Down> to save this information and proceed to the next step.

How to Convert Duties to Tasks

This screen can be accessed by either selecting it from the Background Information Menu, or by striking <Page Down> from the List Duties screen.

```
                            Convert Duties to Tasks                    11-4-96
      Sample Company
      Title - 12345                                           Job Number - 678

      Duty #:      1  Description: _____

          Task #                   Task Description              Task Unit
             1    _____  _____
             2    _____  _____
             3    _____  _____
             4
             5
             6
             7
             8
             9
            10




      F1 Help   F2 Select Duty   F3 Add   F4 Delete   F10 Save
      PgDn Next   PgUp Previous
```

The purpose of the task breakdown step is to break the job being studied down into definable units or "Tasks".

The first duty listed for the job being studied will appear on this screen automatically. You may either start with this duty or select another.

To Select a Different Duty Than the One Displayed

1. Strike the <F2> Key. This will bring up a box with the duties for this job description listed.
2. Use the Arrow Keys to move through the list.
3. Strike the <Enter> key to select. The Duty selected will be displayed in the heading section of this screen.

How to Convert Duties to Tasks

1. Verify the Duty to be broken down into tasks by observing the Duty # and Description displayed at the top of the screen.
2. Type in the tasks that make up this duty in the Task Description column. The form of each task description should be the present tense "ING" form.

```
TASK DESCRIPTION
    Assembling Toy Cars
    Packing Boxes
    Loading Boxes onto Truck
```

3. Strike <Enter>. The cursor automatically moves into the "Task Unit" column.

4.   Type in the Unit of Completion in the "Task Unit" column and strike <Enter>. The cursor will automatically move back into the Task Description column. A unit of completion is the end product of a task. For example, this may be units, units assembled, units moved, or units processed in some way.

TASK DESCRIPTION	UNITS
Assembling Toy Cars	Assembled Cars
Packing Boxes	Boxes Packed
Loading Boxes onto Truck	Boxes

5.   Continue to type in tasks and their corresponding units of completion until all have been entered for this duty.
6.   Strike <F2> to select the next duty to be broken into tasks.
7.   Continue to select duties and enter tasks for the job being analyzed.
8.   Strike <Page Down> to save this information and proceed to the next step.

Getting around on this screen

Getting around on this screen may be accomplished by repetitively striking the <Enter> key. This will move the cursor back and forth and down through the columns displayed. An alternate method is to strike the <Tab> key to move from the right to the left. You may move from left to right by holding the Shift Key and striking the <Tab> Key.

How to Quantify Tasks

This screen can be accessed by either selecting it from the Background Information Menu, or by striking <Page Down> from the *Convert Duties to Task* screen.

This screen pulls together all the information entered thus far and allows you to expand upon. You will determine the Rate, Exposure, and Quantity for each task and then compile a comprehensive list of tasks in the following format:

Task Description
which takes <u>Blank</u> minutes to complete
is performed <u>Blank</u> hours per day
at a rate of <u>Blank</u> units per hour
for a total of <u>Blank</u> units per day.

This screen will present you with all the tasks defined in the previous steps. The duty and task being quantified is indicated by the display in the heading portion of this screen.

```
                         Quantify Tasks

Company - Sample Company                              11-4-96
Title - 12345                                   Job Number - 678
Duty - _____    Task - _____

  Duty:        Unit of      Time Req'd   Hourly    Daily
  Task       Completion       (min)       Rate    Exposure   Quantity
  1:1       _____        0.0         0.0      0.00        0.0
  1:2       _____        0.0         0.0      0.00        0.0
  1:3       _____        0.0         0.0      0.00        0.0
  2:1       _____        0.0         0.0      0.00        0.0
  2:2       _____        0.0         0.0      0.00        0.0
  2:3       _____        0.0         0.0      0.00        0.0
  2:4       _____        0.0         0.0      0.00        0.0
  3:1       _____        0.0         0.0      0.00        0.0
  3:2       _____        0.0         0.0      0.00        0.0



F1 Help  F2 Select Job  F10 Save  PgUp Previous  PgDn Next
Esc Return to Background Menu
```

The columns presented are the:

"Units of completion" as defined in the task breakdown step.
"Time Req'd" for the time required to complete a single unit of completion. (Eg. Time Required to pack a single box, assemble a single widget)
"Hourly Rate" is the number of boxes packed, widgets assembled, or any other units processed per hour.
"Daily Exposure" is the number of hours that this task is performed each day.
"Quantity" is the number of times that a task is performed each day.

You can enter much of the information for this screen in any order you wish. This is often dictated by what information you have available.

1.   Type in the time to complete a single unit in the "Time Req'd" Column. This will be used to fill in the first "blank" statement above. The units in this column are minutes. (If the time required is less than one minute, it must be entered as a decimal number.) Strike <Enter>.
       If you know two out of three of the remaining quantifiers of this task, (rate, exposure, or quantity) the program will calculate the last one.
2.   Enter the Rate in the "Hourly Rate". It will be plugged into the

second "blank" in the sample statement above. Strike <Enter>.

3. Type in the number of hours that a task is performed in the "Daily Exposure" Column. This will result in the immediate calculation and display of the total number of units in the "Quantity" column. The exposure can also be found by entering the total number of units processed per day in the "Quant" column. Entering the Quantity will determine the Exposure for this task only if the rate is entered.

How to Make Changes in Information Entered

Once the various pieces of information has been entered, you can make changes but there are certain considerations.

1. Changing the quantity will alter the hourly rate but not the exposure.
2. Changing the exposure will alter the quantity but not the hourly rate
3. Changing the rate will alter the quantity, but will not affect the exposure.
4. Changing the Time Req'd will not affect anything as this number will simply be plugged into the final statement.

Getting around on this screen

Getting around on this screen may be accomplished by repetitively striking the <Enter> key. This will move the cursor to the right and then down to the next line through the columns displayed. An alternate method is to strike the <Tab> key to move to the right —>. You may move to the left <— by holding the Shift Key and striking the <Tab> Key. You may also use the arrow keys to move up and down a column.

How to Preview, Edit, or Print a Background Report

This function can be accessed by either selecting it from the Main Menu or by striking <Page Down> from the Intervention Discovery Screen.

1. From the "Quantify Tasks: screen. Strike <Escape> to return to the Main Menu
2. Strike <Enter> to save current information (If prompted)
3. Highlight "Print Reports". Strike <Enter>. The "Report Type Selection: menu will display types of reports.

4. Highlight "Background Reports". Strike <Enter>
    The "Background Report Selection" menu will display the options for the different types of Background reports. (See Examples which follow). The heading which indicates the Company name, job title, job number, for which the report will be generated is displayed in the upper section of the screen. Depending on the type of report you will generate, the program will ask for specific information.
5. Highlight the appropriate report format selection. Strike <Enter>
6. Highlight the company name (If prompted). Strike <Enter>
7. Highlight the duty (If prompted). Strike <Enter>
8. Highlight the task (If prompted). Strike <Enter>
9. Accept the default for the report path or type in a different file name and or sub directory. Strike <Enter>
10. Examine and make any changes you like in the report.
11. Strike <F10> to save or <F8> to print.

Completion of these steps will allow you to print out a report in the following format.

Duty List : Presents in a list format the duty #, description, # hours per day, and percent of day as entered in the List Duty screen.

Task Breakdown: Presents in list format the Duty information as above but also lists the associated task numbers and task descriptions as entered in the Convert Duties to Task Screen.

Task Quantification Report - Narrative: Presents in a narrative format a description of each duty, followed by a narrative statement about each task.

Task Description
    which takes <u>Blank</u> minutes to complete
    is performed <u>Blank</u> hours per day
    at a rate of <u>Blank</u> units per hour
    for a total of <u>Blank</u> units per day.

Task Quantification Report - List: Presents task quantification information in a table format. Duty #:Task#, Units of completion, Time to complete, Rate, Daily exposure, and Quantity are listed. *Task description is not included.*

Duty List Report

While performing the job of Toy Maker, the worker performs the following duties:

Duty #	Duty Description	Exposure	Prct
1	Retrieving materials from stock	0.75 Hrs	10.00 %
2	Assembling toys	5.50 Hrs	73.33 %
3	Packing toys	0.50 Hrs	6.67 %
4	Shipping toys out	0.33 Hrs	4.40 %

Duty - Task Breakdown Report

While performing the job of Toy Maker, the worker performs the following duties and tasks:

Duty #	Duty Description	Expo	Prct
1 Retrieving materials from stock		0.75 Hrs	10.00 %
Task # 1	Loading boxes of toy parts from stockroom		
Task # 2	Transporting loaded cart to assembly area		
Task # 3	Unloading boxes at assembly workstation		
2 Assembling toys		5.50 Hrs	73.33 %
Task # 1	Attaching wheels		
Task # 2	Attaching hood		
Task # 3	Painting bottom of toy cars		
Task # 4	Installing headlights		
3 Packing toys		0.50 Hrs	6.67 %
Task # 1	Assembling packing cartons		
Task # 2	Placing cars in boxes		
Task # 3	Sealing cartons		
4 Shipping toys out		0.33 Hrs	4.40 %
Task # 1	Attaching shipping label		
Task # 2	Weighing cartons		
Task # 3	Affixing postage		

Task Narrative Report

Duty #1:   While performing the duty of Retrieving materials from stock, the worker performs the following tasks:

Task #1   Loading boxes of toy parts from stockroom which takes 1.00 minutes to complete is performed 0.75 hours per day, at a rate of 26.7 Boxes per hour, for a total of 20.0 Boxes per day.

Task #2   Transporting loaded cart to assembly area which takes 5.00 minutes to complete is performed 0.75 hours per day, at a rate of 2.7 Loaded carts per hour, for a total of 2.0 Loaded carts per day.

Task #3   Unloading boxes at assembly workstation which takes 0.25 minutes to complete is performed 0.75 hours per day, at a rate of 26.7 Boxes per hour, for a total of 20.0 Boxes per day.

Duty #2:   While performing the duty of Assembling toys, the worker performs the following tasks:

Task #1   Attaching wheels which takes 1.00 minutes to complete is performed 5.50 hours per day, at a rate of 40.0 Wheels per hour, for a total of 220.0 Wheels per day.

Task #2   Attaching hood which takes 0.25 minutes to complete is performed 5.50 hours per day, at a rate of 10.0 Hoods per hour, for a total of 55.0 Hoods per day.

Task #3   Painting bottom of toy cars which takes 1.00 minutes to complete is performed 5.50 hours per day, at a rate of 10.0 Car bottoms per hour, for a total of 55.0 Car bottoms per day.

Task #4   Installing headlights which takes 0.33 minutes to complete is performed 5.50 hours per day, at a rate of 20.0 Headlights per hour, for a total of 110.0 Headlights per day.

Duty #3:   While performing the duty of Packing toys, the worker performs the following tasks:

Task #1   Assembling packing cartons which takes 0.25 minutes to complete is performed 0.50 hours per day, at a rate of 110.0 Cartons per hour, for a total of 55.0 Cartons per day.

Company: Toys Incorporated
Job Title: Toy Maker   Analyst: Mike Burke
Job Number: 88   Date: 04/24/93

Task Narrative Report

Duty #3:   While performing the duty of Packing toys, the worker performs the following tasks (Continued):

Task #2   Placing cars in boxes which takes 0.10 minutes to complete is performed 0.50 hours per day, at a rate of 110.0 Cars per hour, for a total of 55.0 Cars per day.

Task #3   Sealing cartons which takes 0.05 minutes to complete is performed 0.50 hours per day, at a rate of 110.0 Cartons per hour, for a total of 55.0 Cartons per day.

Duty #4:   While performing the duty of Shipping toys out, the worker performs the following tasks:

Task #1   Attaching shipping label which takes 0.16 minutes to complete is performed 0.33 hours per day, at a rate of 166.7 Labels per hour, for a total of 55.0 Labels per day.

Task #2   Weighing cartons which takes 0.50 minutes to complete is performed 0.33 hours per day, at a rate of 166.7 Cartons per hour, for a total of 55.0 Cartons per day.

Task #3   Affixing postage which takes 0.22 minutes to complete is performed 0.33 hours per day, at a rate of 166.7 Cartons per hour, for a total of 55.0 Cartons per day.

Company: Toys Incorporated
Job Title: Toy Maker   Analyst: Mike Burke
Job Number: 88   Date: 04/24/93

Task Quantification Report - List

Dty#:Tsk#	Units	Time mins	Rate units/hour	Expo hour/day	Quantity unit/day
1:1	Boxes	1.0	26.7	0.8	20.00
1:2	Loaded carts	5.0	2.7	0.8	2.00
1:3	Boxes	0.3	26.7	0.8	20.00
2:1	Wheels	1.0	40.0	5.5	220.00
2:2	Hoods	0.3	10.0	5.5	55.00
2:3	Car bottoms	1.0	10.0	5.5	55.00
2:4	Headlights	0.3	20.0	5.5	110.00
3:1	Cartons	0.3	110.0	0.5	55.00
3:2	Cars	0.1	110.0	0.5	55.00
3:3	Cartons	0.1	110.0	0.5	55.00
4:1	Labels	0.2	166.7	0.3	55.00
4:2	Cartons	0.5	166.7	0.3	55.00
4:3	Cartons	0.2	166.7	0.3	55.00

Risk Identification

The Risk Identification step allows you to identify and record specific risk factors associated with each job, duty, or task. A Risk Factor is any condition or practice which may reduce worker comfort and health or act as an obstacle to maximum productivity and consistent quality.

Risk Factors are categorized into five main headings: End Range Positions, Unsupported Posture, Forceful Exertion, Environmental factors, and Excessive Energy Demand.

End Range Position: Moving a joint in the body as far as it will go or close to it.
Unsupported Posture: Anytime a body part is held against gravity for an extended period of time without support.
Forceful Exertion: Forceful exertion deals with the outside agent which must be overcome to perform this step, task, or duty. In most cases, that outside agent is either gravity or friction.

Environmental Factors: Environmental factors take into consideration the way the body interacts with its surroundings and includes such things as heat, cold, vibration and others.

Excessive Energy Demand: Working at a highly demanding pace or level of physical exertion for a prolonged period of time.

How to Access Risk Identification

The Risk Identification screen is accessed by highlighting "Risk Identification" on the main menu and hitting <Enter> or by striking the <Page Down> key from the Task Quantification screen.

Use the Arrow Keys to move the cursor to highlight whether you will be identifying risk factors associated with the entire job, with a specific duty, or with a specific task. The program will prompt you to specify the job title, duty, or task for the risk identification process.

The Risk Identification grid will be displayed. Four of the Risk Factors are displayed along the top of the grid. The various body parts are listed down the left side of the grid. Use the arrow keys to move through the various combinations of risk factors and the body parts where they occur.

```
Sample Company                                          11-4-96
Job Title - 12345                    Job Number - 678
Duty -                               Task -

                End Range      Unsupported    Forceful    Environmental
                Position       Posture        Exertion    Factors
Foot/Ankle
      Knee
       Hip
  Low Back
      Neck
  Shoulder
     Elbow
     Wrist
      Hand

  Factor:  Worker assumes an end range position at the Foot/Ankle
  Detail:                        Quantity/Exposure:    0
Comments:
```

```
F1 Help  F2 Set Job/Duty/Task   F5 Edit Risks  F6 Phys. Risks
F7 Add Comment   F10 Save   PgDn Next   Esc Return to Main Menu
```

The initial format of the risk identification statement is displayed at the bottom of the screen next to the word "Factor:", It will change as you move the cursor around the grid. Because the cursor starts out in the Foot/Ankle box and the End Range box the statement displayed at the bottom is "Assumes as end range position at the Foot/ Ankle".

Getting around on this screen

Moving the cursor up and down in this column by using the arrow up and arrow down keys. The risk identification statement changes on the bottom of the screen.

Moving the cursor to the right or left will result in the risk factor identified in the "factor" statement changing.

Moving the cursor to the up or down will result in the body part identified in the "factor" statement changing.

How to Enter an End Range Position

1.   Highlight the appropriate body part in the "End Range" position of the grid. Strike <Enter>
      A pop up menu will allow you to indicate what end range position is being assumed.
2.   Highlight the appropriate option. Strike <Enter>. A new pop up screen will ask to enter the number of times per task that this is performed. Type in the number and strike <Enter>. A pop up screen will indicate that the risk has been added and prompt you to hit <Esc> to continue. This will return you to the Risk Identification Grid.

How to Enter an Unsupported Position

1.   Highlight the appropriate body part in the "Unsupported Position" position of the grid. Strike <Enter>
      A pop up menu will allow you to indicate what unsupported position is being maintained.
2.   Highlight the appropriate option. Strike <Enter>. A new pop up screen will ask to enter the number of times per task that this is performed. Type in the number and strike <Enter>. A pop up screen will indicate that the risk has been added and prompt you to hit <Esc> to continue. This will return you to the Risk Identification Grid.

How to Enter a Forceful Exertion

1.   Highlight the appropriate body part in the "Forceful Exertion" position of the grid. Strike <Enter>
      A pop up menu will allow you to indicate the type of forceful exertion is being applied.

2. Highlight the appropriate force type option. Strike <Enter>. A pop up screen will ask to enter the amount of force being applied.
3. Type in the number of units times per forceful exertion and strike <Enter>. A pop up screen will ask to enter the units of force being applied.
4. Highlight the appropriate force unit option. Strike <Enter>. A new pop up screen will ask to enter the number of times per task that this is performed. Type in the number and strike <Enter>. A pop up screen will indicate that the risk has been added and prompt you to hit <Esc> to continue. This will return you to the Risk Identification Grid.

How to Enter an Environmental Condition

1. Highlight the appropriate body part in the "Environmental Condition" position of the grid. Strike <Enter>
    A pop up menu will allow you to indicate what environmental condition is being maintained.
2. Highlight the appropriate "environmental" option. Strike <Enter>. A new pop up screen will ask to enter the number of hours per task that this is performed. Type in the number and strike <Enter>. A pop up screen will indicate that the risk has been added and prompt you to hit <Esc> to continue. This will return you to the Risk Identification Grid.

How to Enter an Additional Comment

While you can enter additional comments about a particular risk factor at any time, this is best done immediately after the "Risk Added" pop up screen.

1. Highlight "Additional Comments". Strike <Enter>. A box will open in the Comments section of the screen.
2. Type in the additional comments. Strike <Enter> Strike <Escape> to return to the Risk Identification Grid

How to Enter a Sign of Excessive Physiological Demand

1. Strike <F6> A pop up menu will allow you to indicate what sign of excessive physiological exertion is identified.
2. Move the cursor to the "appropriate "Sign" selection. Strike <Enter>
3. Strike <Escape> or <Enter> to return to the risk identification grid.

How to Enter Other Additional Comments at Any Time
1.   Strike <F7>
2.   Type in the additional comments. Strike <Enter>.

How to Save the Risk Identification Data
1.   Strike <F10> Saving Current Risk Data will be briefly displayed.

How to Review Risk Factors Entered
1.   Strike <F5>. The Risk List Edit screen will appear.
2.   Use the <Page Up> and <Page Down> keys to review each risk factor
3.   Comments may be attached to each risk factor at this time

How to Preview, Edit, or Print a Risk Factor Identification Report
1.   Strike <Escape> to return to Main Menu
2.   Strike <Enter> to save current information
3.   Highlight "Print Reports". Strike <Enter>
     The Report Type Selection menu will display types of report options.

Verify at this time that the company and job title displayed in the heading of this screen is the report you wish to generate. If not, see section below on designating jobs for report writing.

4.   Highlight "Risk Identification Reports". Strike <Enter>
     A pop up menu will display whether the report is to be generated by job title, by duty, or by task.
5.   Highlight the appropriate format selection. Strike <Enter> (See Note below)

Note : You must have identified the risk factors using the format requested. If you try to generate a report by job title after performing the Risk Identification by task, the program will pop up that you must first identify risk factors.

6.   Highlight the company name (If prompted). Strike <Enter>
7.   Highlight the duty (If prompted). Strike <Enter>
8.   Highlight the task (If prompted). Strike <Enter>
9.   Accept the default for the report path. Strike <Enter>

10. Examine and make any changes you like in the report.
11. Strike <F10> to save or <F8> to print.

How to Designate the Job Title for Report Writing
1. Return to "Main Menu"
2. Highlight "Background Information" <Enter>
3. Highlight "Identify Job" <Enter>
4. Strike <Tab> to move the cursor into the 'Jobs Previously Analyzed" box.
5. Highlight the job for which you wish to write a report. <Enter>
6. The "Determine Job Exposure" screen will appear. Strike <Esc>.
7. You will be prompted to save the information for this job. Strike <Enter>
   The "Background Information Menu" will be displayed.
8. Highlight "Return to Main Menu" and strike <Enter> **OR** Strike <Esc>
9. Highlight "Print Reports". Strike <Enter>. Proceed as described above.

Ergonomic Risk Factor Identification (by Job Title)
Ergonomic Risk Factor Identification (by Duty)
Ergonomic Risk Factor Identification (by Task)

Ergonomic Risk Factors Report (by Job Title)

The following is a list of ergonomic risk factors identified for the job title of Toy Maker. An Ergonomic Risk Factor is a condition or practice that can decrease worker comfort, act as an obstacle to maximum productivity and consistent quality, and may increase the incidence of cumulative trauma disorders.

Risk Factor

Worker assumes an end range position at the Foot/Ankle - turns foot inward
   Detail: (inverts)

Worker maintains an unsupported position at the Knee - straight/backbend
   Detail: (hyper extends)

Worker applies a pushing force with the Hip
Detail: Force to be overcome is 25.00 Pounds

The Low Back is exposed to heat or hot surfaces

Worker exhibits signs of excessive physiological demand
Detail: profuse sweating

Ergonomic Risk Factors Report (by Job Duty)

These risk factors were identified while performing the duty of
Retrieving materials from stock which is performed for 10.00 hours.

Risk Factor

Worker assumes an end range position at the Knee - bends knee
Detail: (flexes)

Worker maintains an unsupported position at the Hip - bends
forward
Detail: (flexes)

Worker applies a lifting force with the Low Back
Detail: Force to be overcome is 25.00 Pounds

The Neck is exposed to cold or cold surfaces

Worker exhibits signs of excessive physiological demand
Detail: inability to speak regularly

Ergonomic Risk Factors Report (by Task)

The following is a list of ergonomic risk factors identified for the
job title of Toy Maker. An Ergonomic Risk Factor is a condition or
practice that can decrease worker comfort, act as an obstacle to
maximum productivity and consistent quality, and may increase the
incidence of cumulative trauma disorders.

These risk factors were identified while performing the task of
Attaching wheels, which takes 1.00 minutes to complete is performed 5.50 hours per day at a rate of 40.0 Wheels per hour for a total
of 220.00 Wheels per task.

Risk Factor	Quantity
Worker assumes an end range position at the Hip - turns leg out Detail: (external rotates)	2.00/task
Worker maintains an unsupported position at the Low Back - forward bends Detail: (flexes)	1.00/task
Worker applies a pushing force with the Shoulder Detail: Force to be overcome is 25.00 Pounds	3.00/task
The Elbow is exposed to hard or sharp surfaces	5.50 hours/day
Worker exhibits signs of excessive physiological demand Detail: rubbing or massaging of body part	

Intervention Discovery

The Intervention Discovery step allows you to identify and record specific interventions to help alleviate or reduce the effects of ergonomic risk factors associated with each job, duty, or task.

The best and most effective interventions are the ones that are arrived at independently by using imagination, common sense, and vision. To assist in the creative problem solving process, the Intervention Discovery section of the program will provide several possible solutions to choose from. In order to insure that as many interventions as possible are considered, three different approaches are presented. These approaches are Process Based Interventions, Worker Based Interventions, and Workstation Based Interventions. Each of these approaches can be broken into "Intervention Categories".

Process Intervention Approach
   Input: Changing the raw materials or the way that the INPUT of
   a process is presented to the worker

Output: Changing the output or the speed or level of completion of the product of the worker

Work Station Consideration Approach

Workstation Adjustments: Making changes in the position of the workstation.

Movement Assistance: Providing devices or changes in process to reduce the stress of movement

Assistance to hold in place: Providing devices to help hold something in a more accessible position

Decrease Repetitions: Providing device to help reduce the number of repetitions of certain ergonomic hazards

Worker Support: Providing cushions or supports to allow the worker to rest

Facilitate Access: Changing the work environment to decrease awkward or stressful position

Information Display: Changing the way in which information need to perform a job is presented to the worker

Control Considerations: Changing the way that equipment or devices are controls are positioned or designed

Environmental Controls: Reducing the effects of heat, cold, vibration, hard or sharp surfaces

Maintenance: Providing maintenance programs to keep equipment and devices working a maximum efficiency

Tool Considerations: Possible addition of tools or alternation to existing tools

Worker Involved Program Approach

Training Content Options: What to include in a training program

Training Format Options: Different ways to present the training

Exercise Content Options: What type of exercises to include in an exercise program

Exercise Format Options: Different methods to have the timing and arrangement of exercise programs

Mandate Procedures: Methods of making certain procedures mandatory

Personal Protective Equipment: Options for reducing effects of risk factors by providing the worker with braces, splints, or other protective equipment

Worker Selection: Options for matching a workers ability to the demands of the job

Rotation / Light Duty: Options for the rotation of workers through different jobs or gradually increasing the demands of a job for new or injured workers

> *WHILE THERE ARE SEVERAL INTERVENTIONS LISTED IN THE PROGRAM, IT IS STRONGLY ADVISED THAT YOU TYPE IN AS MANY INTERVENTIONS AS POSSIBLE BY USING THE "EDIT" KEY. THE INTERVENTIONS LISTED HERE ARE MEANT TO ASSIST YOU IN DEVELOPING YOUR OWN IDEAS.*

How to Access the Intervention Discovery Screen

The Intervention Discovery screen is accessed by highlighting "Intervention Discovery" on the main menu and hitting <Enter> or by striking the <Page Down> key from the Risk Identification screen.

1. Use the <Arrow> keys to move the cursor to highlight whether you will be identifying interventions to address the entire job, specific duty, a specific task or a specific risk factor.
2. A series of menus will prompt you to specify the job title, duty, task, or risk factor.

The Intervention Discovery screen will be displayed.

```
                    Intervention Discovery              11-4-96
Company Name: Sample Company
Job Title: 12345                            Job Number: 678
Duty:
Task:
Risk:
                    Select Intervention Category
 PROCESS BASED INTERVENTIONS           Original Process Ideas !
 Input                                 Output
 ********************************       ********************************

                    Select Intervention
 [ ]   Increase/decrease the size of the container/item
 [ ]   Change the position of the container/item
 [ ]   Increase/decrease the weight of the container/item
 [ ]   Change the shape of input
 [ ]   Have materials delivered to a different area
 [ ]   Have the supplier perform part of the task

 F1 Help  F2 Intervention Type  F7 Add General Comment  F10 Save
 Tab Select Interventions  PgDn Next Screen  Esc  Main Menu
```

The screen contains two windows within it. They are the "Select Intervention Category" window and the "Select Intervention" window. Arrow Keys are used to scroll through the categories. Examples of interventions for that category are simultaneously displayed in the "Select Intervention" window.

This part of the program allows you to compose and generate a list of interventions to help reduce or alleviate risk factors in a workstation. This is accomplished by choosing from a list of interventions or adding your own additional comments.

How to Move the Cursor from the "Select Intervention Category" Window to the "Select Intervention" Window

1. Strike <Tab>

How to Move the Cursor from the "Select Intervention" Window to the "Select Intervention Category" Window

1. Strike <Tab>. A pop up menu will ask if you want to save your work.
2. To include the interventions you have just selected in your report, highlight "Yes" and strike <Enter>
3. To ignore interventions you have just selected (or if none selected) highlight "No" and strike <Enter>
4. To return to the Select Intervention window highlight "Cancel" and strike <Enter>

How to Select an Intervention to be Included in the Report

1. Highlight the appropriate category in the "Select Intervention Category" window. Strike the <Enter> key . This will move the cursor into the "Select Intervention" window.
2. Highlight the appropriate intervention. Strike <Spacebar>. This will place a check mark next to the intervention selected. (Hitting the <Spacebar> again will remove the check mark.)
3. Use the Arrow Keys to highlight any other interventions in this category and repeat instructions in number 2 above.
4. When all appropriate interventions have been selected, strike <Tab> to return to the intervention category window. A pop up menu will ask if you want to save your work.
5. To include the interventions you have just selected in your report, highlight "Yes" and strike <Enter>
6. To ignore interventions you have just selected (or if none selected) highlight "No" and strike <Enter>
7. To return to the Select Intervention window highlight "Cancel" and strike <Enter>

How to Add Additional Comments or Expand upon a Listed Intervention

1. Strike <Enter> immediately after you have selected that intervention and the check mark is displayed. A blank text window will pop up.
2. Type the comment. Strike <Enter>.

How to Add Your Own Intervention Ideas

1. Position the cursor anywhere in the "Select Interventions Category" window.
2. Strike <F7>. A blank text window will pop up.
3. Type the comment. Strike <Enter>.

How to Start a New List of Interventions for a Different Job Title, Duty, Task, or Risk Factor

1. Strike <F2>. A series of menus will prompt you to specify the job title, duty, task, or risk factor.
2. Highlight and select the appropriate choices.

Do not hesitate to enter ideas which seem radical or totally impractical. These impractical ideas often lead to a discovery of a more practical and effective suggestion.

Some Special Considerations

How to Enter a Training Program

1. From the Select Interventions Category menu, highlight "Training Format Options". Strike <Enter>
2. From the "Select Intervention" menu, highlight the appropriate training format selection. Strike <Spacebar> A check mark will appear next to your selection.
3. Immediately strike <F7>. The "Edit Intervention Comment" box will be displayed.
4. Type in the statement, "Program should contain the following components:" Strike <Enter>
5. Strike <Tab> to move back into the Select Intervention Category window.
6. Strike <Enter> to save your work.
7. Immediately striike the <Right Arrow > key to highlight "Training ContentOptions". Strike <TAB>
8. Use the Arrow keys to highlight appropriate components for inclusion in your suggested training session. Strike <Spacebar> to select.

9. Strike <Tab> to return to "Select Intervention Category" when you are done.
10. To add any components not listed:
    a. Strike <F7> The "Add Miscellaneous Intervention Comment" box will be displayed.
    b. Type in component you wish to add to training session. Strike <Enter>
    c. Repeat this process to add as many additional training content components as you which

How to Enter an Exercise Program

1. Highlight "Exercise Content Options" from the "Select Intervention Category" window. Strike <Tab>
2. Highlight the statement that best describes your recommendation for an exercise program. Strike <Spacebar> to select.
3. Strike <Tab> A pop up window will prompt you to save your work.
4. Strike <Enter> with the "Yes" highlighted .
5. Strike the Right arrow key once to highlight "Exercise Format Options" from the "Select Intervention Category" window. Strike <Tab>
6. Highlight the statement or statements that best describes your recommendation for the frequency of the exercise program. Strike <Spacebar>
7. Strike <Tab>. A pop up window will prompt you to save your work. Strike <Enter> with the "Yes" highlighted .

Completion of these steps will allow you to print out a report in the following formats.

Intervention List (by Job Title)
Intervention List (by Duty)
Intervention List (by Task)
Intervention List (by Risk Factor)

How to Preview, Edit, or Print an Intervention List Report

1. From the Intervention Discovery screen, strike <Page Down>. The "Saving Data" message will be displayed . The "Report Type Selection" menu will display types of report options. Verify at this time that the company and job title displayed in the heading of this screen is the report you wish to generate.
2. Highlight "Intervention Discovery Reports". Strike <Enter>

A pop up menu will display whether the report is to be generated by job title, by duty, or by task.

3.   Highlight the appropriate format selection. Strike <Enter>

Note: You must have performed the intervention discovery process in the format requested.

4.   Highlight the duty (If prompted). Strike <Enter>
5.   Highlight the task (If prompted). Strike <Enter>
6.   Accept the default for the report path or type in your own. Strike <Enter>
7.   Examine and make any changes you like in the report.
8.   Strike <F10> to save or <F8> to print.

Intervention Report

The following suggestions represent a comprehensive list of actions which may be taken to provide a more effective work environment. Implementation of these interventions will help to increase worker comfort level, facilitate maximum productivity and quality, and reduce the incidence of many cumulative trauma disorders. Determination of the most appropriate steps will depend on the unique circumstances in this facility.

Job Title: *(As Indicated)*
Duty: *(As Indicated)*
Task: *(As Indicated)*
Risk: *(As Indicated)*

Attach work surfaces to wall to allow variability in desk height

Install shelves over workstations to allow storage of less frequently used items and thus free up work space

Install split level work surface heights to allow placement of monitors at one level and the keyboards at another level

Install adjustable workstation

Use of personal braces or splints to limit motion

Provide guidelines for adjustment of chairs

Provide adjustable footrest

Provide adjustable armrests that attach to workstation

Bevel all workstation edges and railing

Provide alternate seating devices such as ergodyne or balans chairs

Provide personal heating or cooling devices

Initiate live interactive training program

Training program to include
  Exercises to perform at work
  Least stressful work practices

PRINT REPORTS

Upon completion of the analysis steps you will want to present the information in a logical and easy to understand format. This program provides a model for the compilation of a final report. While you are free to select any combination of report to suit your needs, the following model may provide a good basis for development:

> Introductory Statement
> Background Information Reports
> Risk Identification
> Intervention Discovery
> Summary and Recommendations

You have the option of editing reports generated in this section using this program or saving it to a file and then editing it in a word processing program.

How to Access the Print Reports Section from the Main Menu

1.   Highlight "Print Reports" on the Main Menu. Strike <Enter>. The "Report Type Selection" menu will appear.

How to Access Print Report Section from the Intervention Discovery Section

1.   When all the intervention have been entered and saved, Strike <Page Down>. The "Report Type Selection" menu will appear.

How to Generate a Specific Type of Report

1. From the "Report Type Selection" menu, use the Arrow keys to highlight the type of report you wish to create. Strike <Enter>
2. The additional choices for the type of report you wish to generate will be displayed. Continue to highlight and strike <Enter> until you are prompted to save the data.
3. You may either accept the default name for the report or create your own. Accepting the default will overwrite the last report of the type selected.

It is strongly advised that you rename the file at this time in a way which will allow you to distinguish it from any other analyses performed! You will not be given the opportunity to change the name from within this program after this.

4. Type in a report file name or accept the default and strike <Enter>

Most of the reports have been illustrated in earlier parts of this manual. The exception is the Introductory statements and Summary and recommendations. The format for those are going to vary greatly depending on the unique circumstances surrounding the use of the program. Samples are included as a model only.

How to Generate a Complete Report Made Up of Several Report Types

1. From the "Report Type Selection" menu, highlight "Generate Complete Report". Strike <Enter>. The Ergonomics Analysis - Complete Report Options" menu will be displayed.
2. Use the Up and Down Arrow Keys to move the cursor throughout the options. Strike <Space> to select or de-select various reports to be included in the complete report.

You must have performed the specific step in the ergonomics analysis using the format that you highlight .

How to Write an Introductory Report

The introductory statement should contain the date of the analysis, the name of the company, the name of the job, and perhaps even the reason for the analysis.

The following example may be helpful.

> On October 22, 1991, an ergonomic analysis was performed at A.V.B. Inc. located at 753 Industrial Drive, Bronx, New York. The major product produced at this facility is industrial training aids.
>
> A.V.B. Inc employees 350 persons working two shifts. In the past three years there has been a change in facility management. In addition, lift belts and back injury prevention training has been initiated on a pilot program. No objective data is available regarding the relative success of these interventions.
>
> The job title that was analyzed was the Order Picker and Packer. 78 persons are employed to perform this job function.

You may also choose to type in a narrative description of the job. This is not also necessary it does help to indicate the analysts understanding of the basic job functions.

> Order Pickers/packers are responsible for pulling materials from storage shelves, packing the various pieces into boxes, addressing boxes, and transporting to the postage and shipping area.
>
> Order Pickers/packers work 5 days per week. They are given two paid 15 minute coffee breaks and one unpaid 45 minute meal break. The total job exposure is 7 1/2 hour per day.

1. From the "Report Type Selection" menu, highlight "Introduction Report" and strike <Enter>. The report path screen will be displayed.
2. Accept the default presented or type in your own directory, sub directory, and filename designation. Strike <Enter>.

It is strongly advised that you rename the file at this time in a way which will allow you to distinguish it from any other analyses performed! You will not be given the opportunity to change the name from within this program after this.

3. If you accept the default and have ever written a report using this designation before, you will be asked to approve overwriting that report.
4. Highlight "Yes" and strike <Enter> to start a new report.

Highlight "No" and strike <Enter> to continue to work on a previously written report.

5. Enter the text of the report. The word processing functions of the program are limited . You will be able to import this file into most word processing programs to improve the overall appearance.
6. Strike <F10> to save or <F8> to print.

How to Generate a Summary and Recommendations Report

The Summary statement is an opportunity to point out the most important findings and recommendations for immediate initiation of interventions. This should not be longer than one page although you may choose to refer to sections within other sections of the report. The following example may act as a short model of the type of information included in a summary and recommendations report.

Summary and Recommendations

The job of Toy Maker was identified by Mr Smith, the Health and Safety Coordinator, for ergonomic analysis. This was as a result of several recent low back injuries and a shoulder injury. Following this analysis, several controllable risk factors which may affect worker performance, productivity, comfort, and safety were identified.

Following a duty task breakdown, the primary task was identified as attaching wheels. This task is performed the greatest part of the day. The greatest number of risk factors were associated with this assembly process.

Of all the interventions mentioned above, the most appropriate would be initiation of an education program, initiation of a tool maintenance program, installation of a tool balancer, providing cushioning on the edges of the work station, and reorganizing shelves to put least called for items on the higher shelves.

How to Generate a Report About a Previously Analyzed Job

1. From the Main Menu, select "Background Information"
2. From the Background Information menu, select "Identify Job"
3. From the "Identify Job screen, strike <Tab> to move into the "Jobs Previously Analyzed" window.

4. Highlight the appropriate job title. Strike <Enter>. The job exposure screen will be displayed.
5. From the "Job Exposure" screen, strike <Esc>
6. Strike <Enter> to accept the question to save the current job exposure information.
7. Highlight the "Return the Main menu" Strike <Enter>
8. Highlight "Print Reports" Strike <Enter>
9. Proceed as described in "Background Information, Risk Identification, or Intervention Discovery section of this manual.

How to Import the Files Into a Word Processing Program

The word processing options of this program are limited. This was done to help keep the cost of the program down. All the reports are saved as ASCII text files. You can generate them and name them in a way that will facilitate recognition using what ever extension you like. When the program prompts you to confirm the Report path, change the name and the sub directory to save the report to: The name can be up to 8 letters long. Most word processing programs will allow you to import these files and manipulate within that environment.

How to Exit Program

To exit the program, strike the escape key to return to the main menu. Highlight the "Exit " option and strike enter.

How to Modify the Utilities Section

When you installed the program you were prompted to enter your name and the various report paths for saving Data and reports. You may want to modify that at some time to facilitate saving reports to other sub directories without having to enter that information each time you generate a report. This should only be attempted by experienced computer users.

Access to the Utilities Screen is by highlighting Utilities from the main menu <Enter>.

T - #0238 - 101024 - C0 - 216/138/3 [5] - CB - 9781315892795 - Gloss Lamination